SCIENCE FOR KIDS

Earth

Our Created Home

SCIENCE FOR KIDS

Earth

Our Created Home

Illustrated by Susan Windsor

INSTITUTE FOR CREATION RESEARCH

Dallas, Texas
ICR.org

SCIENCE FOR KIDS

Earth
Our Created Home

First printing: August 2019

All Scripture quotations are from the New King James Version.

ISBN: 978-1-946246-32-5
Library of Congress Catalog Number: 2019907849

Please visit our website for other books and resources: ICR.org

Printed in the United States of America.

Table of Contents

Earth: Our Home

- *Some scientists think Earth formed by accident over billions of years.*
- *The Bible says God created Earth thousands of years ago.*
- *God gave Earth special features no other planet has.*

Where in the world are we? No matter our zip code, the big answer for all of us is the same: Earth! This spinning ball of wonder is far more than just the third rock from the sun. Earth is unlike any other planet anyone has ever discovered.

Many scientists claim that Earth's unique features developed by random chance over billions of years. And

Day 1: Light

Day 2: Waters

Day 3: Dry Land

they think our presence in this world is just another happy accident. But the Bible sheds light on the real story.

In the beginning God created the heavens and the earth. The earth was without form, and void; and darkness was on the face of the deep. And the Spirit of God was hovering over the face of the waters. (Genesis 1:1-2)

God spent six days creating the entire universe, from far-off galaxies to the leaves on our trees. He created plants and animals to fill the land, sky, and sea. He also made the first people, Adam and Eve. And according to the Bible's timeline, all of this happened about 6,000 years ago.

Day 4: Sun, Moon, and Stars	Day 5: Sea Animals and Birds	Day 6: Land Animals and Humans

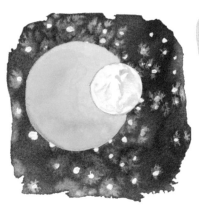  

The Fall

- *God made all of creation "very good."*
- *God cursed the world because of Adam's sin.*
- *This is why Earth has thorns, thistles, and other harmful things.*

Have you ever wondered why a world filled with so much goodness and beauty is also filled with sorrow and brokenness? Why are the soft petals of a rose wrapped so delicately while its stem is covered in prickly thorns?

When God made everything during the creation week, He declared it "very good." But Adam and Eve chose to disobey God's law in His very good world. Because of their sin, He cursed His beloved creation.

> "Cursed is the ground for your sake; in toil you shall eat of it all the days of your life. Both thorns and thistles it shall bring forth for you." (Genesis 3:17-18)

This explains why so many marvelous wonders of Earth are entangled with thorns and thistles, death and disease. Earth is groaning, waiting eagerly for God to make it new (Romans 8:22). At the right time, He will do just that.

Earth and the Bible

- *The Bible says that God made Earth.*
- *God keeps Earth in existence.*
- *Someday He will make our planet new.*

What can the Bible tell us about Earth? More than you might think!

God made Earth. "Ah, LORD God! Behold, You have made the heavens and the earth by Your great power and outstretched arm. There is nothing too hard for You" (Jeremiah 32:17).

God owns Earth. "The earth is the LORD's, and all its fullness, the world and those who dwell therein" (Psalm 24:1).

Earth moves in a circular orbit. "It is [God] who sits above the circle of the earth, and its inhabitants are like grasshoppers" (Isaiah 40:22).

God keeps Earth from going out of existence. "[God] laid the foundations of the earth, so that it should not be moved forever" (Psalm 104:5).

God will make Earth new again. "For behold, I create new heavens and a new earth; and the former shall not be remembered or come to mind" (Isaiah 65:17).

Earth: Inside and Out

- *Earth has four layers: the crust, mantle, outer core, and inner core.*
- *Earth is mainly made of rock and possibly metal.*
- *God formed Earth at the beginning of creation.*

Earth rocks, but it isn't just one big hunk of rock. Our planet has four layers, and each layer has its own special features. We live on Earth's outer layer—the **crust**. It's mostly made of granite and basalt rock. The next layer, the **mantle**, is made of iron and magnesium-rich rocks. It's nearly all solid.

Super-hot liquid, possibly made of iron and nickel, flows through the **outer core**. The **inner core** lies at Earth's center. It's probably made of very hot iron and nickel too, but high pressure keeps these metals in solid form. The Bible says that God formed Earth at the beginning of creation (Genesis 1:1), and these four layers are part of His excellent design.

Did you know? Earth's magnetic field is an invisible shield that reaches into outer space and protects us from harmful solar winds. Some scientists believe that the movement of metals in Earth's outer core create the magnetic field, but none of them know how that could work. God probably created Earth's magnetic field during the creation week.

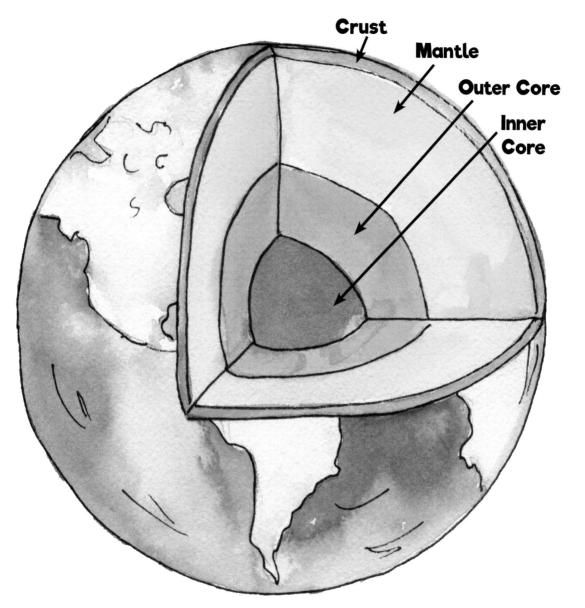

Crust

Mantle

Outer Core

Inner Core

Geography Lesson

- *Earth's crust has seven land masses called continents.*
- *The continents are surrounded by five oceans.*
- *A big event in history formed many of the geological features we see today.*

To get to know our home planet a little better, let's start with some geography basics. Earth's crust has seven massive pieces of land called continents: North America, South America, Europe, Asia, Africa, Australia, and Antarctica. Natural barriers like oceans or mountain ranges separate each continent from the others. The continents display a wide range of weather, geological features, plants, and living creatures.

Water covers the rest of the planet, and it's divided into five major oceans: the Pacific Ocean, Atlantic Ocean, Indian Ocean, Arctic Ocean, and Southern Ocean.

The Bible says that God made land and water during the creation week. But many features of Earth's continents and oceans formed during a disastrous event in history—the global Flood. Keep reading to discover how God's response to people's sinful choices changed our planet forever.

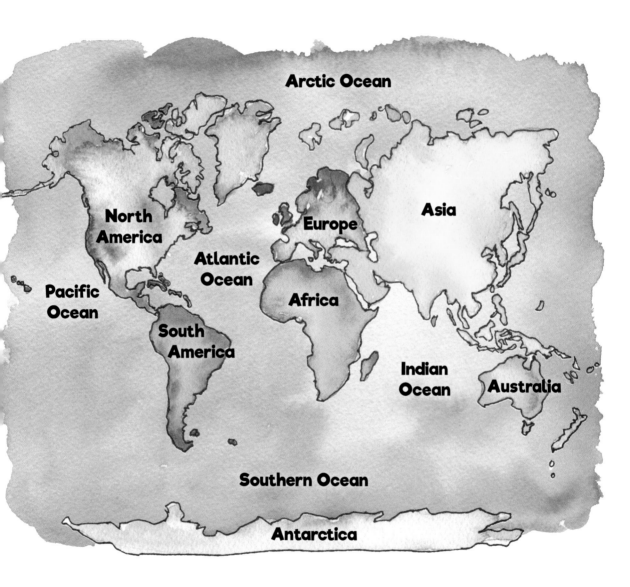

Life Is Beautiful

- *God created Earth to be filled with life.*
- *Earth is full of people and animals.*
- *God has given us everything we need to live on this planet.*

On planet Earth, over seven billion people live, play, work, and worship. Six of the seven continents are home to people of various colors, shapes, and sizes—each with their own traditions and culture. While a few brave, hardy humans visit Antarctica, no one lives there permanently.

Scientists estimate that about 6.5 million species—like horses, chickens, and crickets—live on land. And about 2.2 million species—including lobsters, coral, and whales—dwell in Earth's oceans. Is it by accident that our planet overflows with life? Not at all! An earth full of animals and people is exactly what God had in mind from the beginning.

> Then God blessed [Adam and Eve],…"Be fruitful and multiply; fill the earth and subdue it; have dominion over…every living thing that moves on the earth." (Genesis 1:28)

Our world is marked by sin. Even so, God has given people and animals everything we need to live out His purpose for us.

Peace Again

- *God created people and animals to live peacefully on Earth.*
- *When sin entered the world, people and animals began to eat each other.*
- *When God remakes our planet, there will be peace again.*

In the beginning, Earth's animals and people lived together in peace. All ate plants (Genesis 1:29-30), and none ate each other.

But when sin entered the world through Adam, our planet became violent. Teeth and claws were originally designed for chomping veggies and digging roots, but animals began to use them for tearing and eating flesh. God later granted people permission to eat animal meat (Genesis 9:3).

One day, when God makes Earth new again, peace will return. The Bible gives us a wonderful picture of this future:

> "The wolf also shall dwell with the lamb, the leopard shall lie down with the young goat, the calf and the young lion and the fatling together; and a little child shall lead them." (Isaiah 11:6)

Did you know? Panda bears and fruit bats use their sharp teeth to eat plants, not meat. This is how it was with all sharp-toothed creatures at the beginning of creation.

Our Round Earth

- *The Greeks figured out that Earth is a sphere, or ball.*
- *Earth casts a round shadow on the moon during an eclipse.*
- *Christopher Columbus didn't discover Earth's shape.*

We have beautiful pictures of our round Earth taken from space. But what did people know about Earth's shape before space travel?

About 2,500 years ago, ancient Greek philosophers figured out the earth is round. What were their clues? They thought that since the moon is round, maybe Earth is too. Earth casts a round shadow on the moon during a lunar eclipse. They also noticed that ships sailing toward the horizon disappear from the bottom up. And people in other locations have different views of the stars.

You may have heard that Christopher Columbus proved the earth is round. But that's a myth. People already knew Earth's shape. What many of them didn't know about was the American continents.

Did you know? When Isaiah 40:22 mentions the "circle of the Earth," it's referring to our planet's movement in a nearly circular orbit around the sun rather than its shape.

**Earth's shadow
on the moon**

Mapping Earth

- *Earlier people didn't know much about other parts of the world.*
- *Explorers and mapmakers helped them learn more.*
- *Now, satellites provide detailed images of our planet.*

The Greeks started drawing maps thousands of years ago. They used these drawings to navigate the Mediterranean Sea.

Romans and Arabs later found the Indian Ocean, distant China, and the deep parts of Africa. At the same time, the Chinese made detailed maps of the areas they knew. When people from the East and the West met, they put their maps together and got a better understanding of the world.

The Americas and their native peoples were mostly unknown to the rest of the world until the Vikings sailed across the ocean in search of new land. Later explorers like Christopher Columbus and Amerigo Vespucci mapped the "new" continents' features.

It took thousands of years to get a good picture of Earth's land and oceans. Today, we can turn on a computer and look at the whole world in just a few seconds.

A Worldwide Puzzle

- *Earth's land likely started out as one huge continent.*
- *Enormous slabs of rock (plates) moved and changed Earth's surface.*
- *Plate movements ripped the one continent into seven.*

Do you like putting puzzles together? Well, about 100 years ago, German meteorologist Alfred Wegener noticed Earth has a puzzle of its own. He saw how the eastern edge of South America and the western edge of Africa could fit together like two disconnected pieces.

Were those two continents once connected to each other and the other five as part of one huge continent? This would match what the Bible says in Genesis 1:9.

Wegener also thought Earth's crust had moved and shifted. He believed this movement is what split the continents apart in the first place. Other scientists thought Wegener's ideas were pretty crazy. But Wegener's theory— now called plate tectonics—is widely accepted today. What major event in Earth's history could have ripped the land into pieces? Turn the page to find out.

Then God said, "Let the waters under the heavens
be gathered together into one place, and let the
dry land appear"; and it was so. (Genesis 1:9)

Global Flood

- *Genesis tells us about the worldwide Flood.*
- *The Ark saved Noah, his family, and many animals.*
- *The Flood changed Earth in big ways!*

About 4,500 years ago, Earth was filled with violence. God warned Noah that He was sending a flood to judge the earth and told him to build a huge boat. When the Flood came, Noah and his family got on board. At least two of each kind of land animal and bird did, too.

The boat was called an ark. It protected all who were inside it for about a year. Then the floodwaters went down, and the Ark landed in the mountains of Ararat. When Noah and his family got out, they were the only humans alive.

The global Flood changed our planet a lot. Huge, moving slabs of crust and underground rock—called tectonic plates—broke the land into separate continents during the Flood. Where the plates slammed together, they pushed up mountains. The Flood's waves, full of sand, mud, and water, buried plants and animals to form fossils. And the layers of sand and mud hardened into rock. Scientists now study these layers to dig up fossils and find clues about Earth's history.

Flood Evidence

- *Evidence of Noah's Flood is everywhere.*
- *It takes a lot of water and mud to bury huge creatures.*
- *Cultures around the world tell stories of an ancient flood.*

The Bible describes a worldwide flood in the book of Genesis. But is there any evidence that this major event ever happened? Actually, evidence is all over the earth!

We find gigantic creatures—like dinosaurs—fossilized on every continent. It would take a lot of mud and water to cover such big beasts. And they'd have to be buried quickly to turn into fossils.

Huge layers of sand and mud covered entire continents.

Most scientists agree that these layers were put there by water. Now, they have hardened into rock. Layers in canyon walls or in roadcuts in the hills and mountains show us evidence of the global Flood.

Cultures all over the world have told stories of a worldwide flood to their children and grandchildren. The details of the story may vary, but the main events are a lot alike. How could people all over the world know about the same sad event unless it really happened?

Did you know? People have found fossils of sea creatures on mountaintops. The global Flood explains these strange discoveries.

Fossils and the Flood

- *Plants and animals must be buried quickly and deeply to become fossils.*
- *The global Flood buried plants and animals all over the earth.*
- *Evidence shows that most fossils formed in the global Flood.*

Plants and animals can only turn into fossils if they are buried quickly and deeply by mud and other sediments. Otherwise, when they die they would rot and disappear, which is what usually happens. What kind of event could bury creatures on every continent, including gigantic dinosaurs? What kind of event could mix land animals with sea creatures? The global Flood described in Genesis could!

About 10,000 duck-bill dinosaurs were found buried together in Montana. What could have covered so many in mud so quickly? A huge flood could. Dead jellyfish rot in a very short time, yet their fossils are found around the world. They had to be buried very fast.

Many fossils are found in a death pose with their necks arched back. This would happen if they died choking on mud. And fossils of clams are often found tightly shut. Clams close their shells when they sense danger. These discoveries make perfect sense because the Bible says the Flood covered the whole earth.

What Fossils Tell Us

- *Fossils show instant creation, not evolution.*
- *Fossils show that animals' basic body styles have stayed the same over time.*
- *Fossils show some creature kinds that are now extinct.*

What else can Earth's fossils tell us? Fossils show that each creature kind has the same basic look no matter where it appears in Earth's rock layers. Clams look like clams. Frogs look like frogs. Though some may vary in certain traits, the fossil record shows little change from the basic body styles we see today.

Evolution scientists believe that the farther down in Earth's rock layers you go, the further back in evolution's timeline you go. They expect simple creatures at the bottom and different, complex creatures in upper layers. But fossils don't show evolution. Even creatures found in the lowest rock layers, like trilobites, have very complex features.

Fossils also tell us that basic types of plants and animals existed at the same time. And each fossil creature shows up with all of its parts, just as if God made it. Earth's rock layers full of fossils are exactly what we would expect to see if a past flood buried God's creatures all over the world.

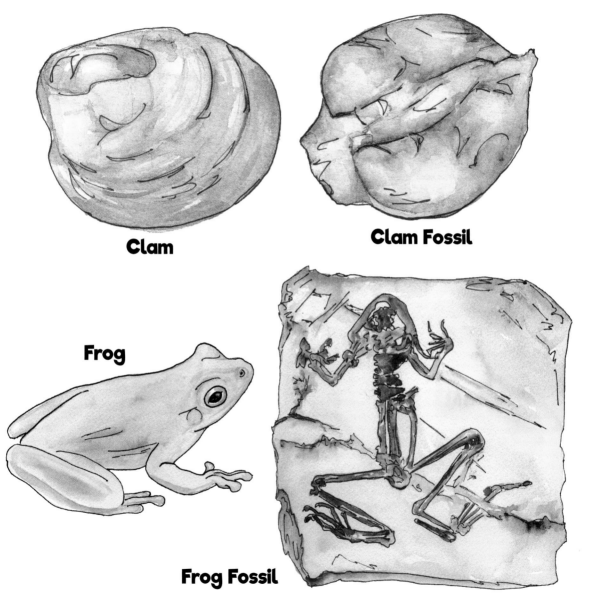

Clam

Clam Fossil

Frog

Frog Fossil

Soft Tissue Surprises

- *Soft tissues like skin, blood vessels, and bone cells can't last for millions of years.*
- *Soft tissues have been discovered in dinosaur and other fossils.*
- *Fossils must be younger than many scientists think.*

Have you ever seen a dead fish on the beach or a dead animal on the side of the road? Those sorts of things don't last very long. Bacteria and other creatures eat away at them until they are completely gone.

The creatures in the fossil record would have been eaten, too, if they hadn't been buried so quickly. But even buried beneath rock, their soft tissues—like skin, blood vessels, and bone cells—can't last for millions of years. So, why have so many of these tissues been found in fossils?

Blood vessels and skin cells were found in duck-bill dinosaurs. A *Triceratops* had soft tissue in its horn. And a squid fossil contained ink that was so fresh scientists used it to draw a picture! Some say that the soft tissues aren't *really* soft tissues. But every experiment confirms that the tissues are real. These creatures must have been alive more recently than many scientists think.

Fossil Fuels

- *The global Flood buried plants and animals.*
- *High heat underground turned some of them into oil and coal.*
- *Oil and coal form in a short time, not over millions of years.*

Fossil fuels like oil and coal provide energy for our world. They power cars, trucks, trains, and factories. You may have heard it takes millions of years for them to form—but is that true?

About 4,500 years ago, the global Flood buried tons of tiny sea creatures called algae. One of the keys to making oil is fast burial—before the algae decay. High-pressure heat deep in the ground turned that algae into oil after only a couple hundred years. In a laboratory, scientists turned algae into oil in only one hour!

Coal has a similar story. It formed when certain plants buried during the global Flood endured high heat underground. Coal can be made in a laboratory in only a matter of hours or days—it doesn't require long ages either.

Did you know? Coal often has land and sea creature fossils buried with it—another sign that it formed because of the Flood.

Coal

Grand Canyon

- *Grand Canyon is a deep, wide hole in the western United States.*
- *The global Flood laid down the canyon's colorful rock layers.*
- *Later, the Flood's draining water rapidly carved the canyon.*

Near the mountains of the western United States lies one of the biggest, most beautiful canyons in the world: Grand Canyon. Where did its colorful rock layers come from? You may have heard that they formed over millions of years.

But each rock layer shows no signs that it was changed by wind and water before the next layer formed. Instead, they look like flat layers put down one after the other in a short period of time. The global Flood dropped thick layers like these all over the world, including at Grand Canyon.

Some scientists think the Colorado River slowly carved Grand Canyon over millions of years. But the Colorado River would have had to flow uphill in some places to do this. That's not possible. ICR geologist Dr. Tim Clarey has a more sensible explanation. He thinks a large amount of water from the end of Noah's Flood carved the canyon quickly as it drained west off the Colorado Plateau.

Did you know? Native Americans have lived in and around Grand Canyon for thousands of years.

Before the Flood

- *Science and Bible clues reveal what Earth may have been like before the Flood.*
- *Flood events caused changes in Earth's continents, water systems, and creatures.*
- *We may never know much about the pre-Flood world.*

How was pre-Flood Earth different from today?

One land. Before the Flood, all of the continents were likely joined together into one big supercontinent.

Larger animals. Flood fossils reveal dragonflies with wingspans of up to two feet and a guinea pig the size of a buffalo!

Different water cycle. Genesis tells us that "a mist went up from the earth and watered the whole face of the ground" (Genesis 2:6). Perhaps heat from deep underground raised water into ancient land-misting and river systems that were destroyed by the Flood.

Shallow seas and many swamps. Earth's rocks contain many shallow marine fossils. There were likely many shallow seas before the Flood. And fossils buried with dinosaurs indicate that pre-Flood Earth must have had large swamp-like regions.

Scattering from Babel

- *After the Flood, Noah's descendants built a city called Babel.*
- *Because of their pride, God confused their languages.*
- *The people scattered all over the world.*

Soon after the global Flood, Noah's family multiplied. His three sons—Shem, Ham, and Japheth—and their wives had many children, and those children had children. These are called descendants. They lived in a city called Babel and spoke the same language.

But the people got prideful. They started building a tall tower to prove their greatness. God confused their languages so they couldn't talk to each other. The families of Babel gave up their project and scattered across the world.

Genesis 10 indicates that the descendants of Shem went into the Middle East. The descendants of Ham went into Africa and Asia. The descendants of Japheth went into Europe. This is why people with similar languages and physical features tend to live in the same parts of the world today.

Ice Age

- *The Global Flood caused the Ice Age.*
- *People and animals struggled to live during this time.*
- *Earth shows much evidence that an ice age happened.*

After the worldwide Flood, Earth experienced the Ice Age. The climate was very cold, and ice covered much of our planet for around 700 years. Wooly mammoths, saber-toothed cats, and other creatures roamed during this time. Some animals struggled to survive the extreme temperatures. It was probably hard for people, too.

The global Flood caused the Ice Age. During and after the Flood, the plates of crust and rock beneath Earth's surface moved around. The movement caused volcanoes to erupt, sending dust and ash into the air. This blocked some of Earth's sunlight. Lots of snow fell, and the summers weren't warm enough to melt it.

How do we know the Ice Age happened? For one thing, there's evidence of past glaciers in Earth's Northern Hemisphere. Glaciers are thick sheets of ice that cover the land. Though the Bible doesn't talk about the Ice Age, the book of Job mentions cold, snow, ice, and frost more than any other book in the Bible. Job probably lived during the Ice Age.

Land Bridges

- *During the Ice Age, strips of dry land connected the continents.*
- *Animals and people walked to other continents on these land bridges.*
- *Land bridges helped people to obey God's command to "fill the earth."*

After the Flood, Noah's descendants tried to stay in the city of Babel. They didn't want to spread out and "fill the earth" as God commanded (Genesis 9:1). God divided and scattered the people by giving them different languages. But how did they get across the oceans to other lands, like North America and Indonesia?

Well, when the Ice Age began, a lot of ocean water froze into ice. Sea levels dropped so low that long pieces of land between the continents appeared. These land strips were like bridges that people and animals could walk on.

One land bridge connected Asia and Alaska. Another connected Asia and Indonesia. The land bridges were only there for a few hundred years. After that, the ice melted, sea levels rose, and water covered them once again. When God scattered the people at Babel, they probably crossed the land bridges and were able to fill the earth as He commanded. God's timing is perfect.

Ancient Cities and Writing

- *The Middle East has some of the oldest cities and writing records.*
- *After the Tower of Babel, people scattered from the Middle East and built new cities.*
- *Artifacts from those ancient cities confirm Bible history.*

The Bible says all peoples came from the Middle East (Genesis 11:1-9). Did our ancestors leave evidence to back this up? Some of the oldest civilizations on Earth were in Babylon—a place in the Middle East we call Iraq today. Historians may even have pinpointed the plot of land in Iraq where the Tower of Babel once stood.

Abraham was born around 2000 BC in a city named Ur near Babylon. Some of the earliest examples of writing show a symbol system used in Babylon called cuneiform (kyoo-nee-UH-form). The first evidence of bread, wine, and agriculture came from this region in the Middle East. And the oldest temple in Gobekli Tepe, Turkey, shows that the earliest worshipers came from the Middle East, too.

Some scholars insist that people evolved from ape-like creatures in Africa. But where do we find the oldest city ruins and first writing records? We find them not in Africa but in the Middle East like the Bible says. The oldest civilizations and writings match the Bible's times, peoples, and places.

Cuneiform

Venting Volcanoes

- *A volcano releases hot lava, rocks, dust, and gas from deep within the earth.*
- *Volcanoes can trigger earthquakes and tsunamis.*
- *Volcanoes erupted for many years after the global Flood.*

Have you ever shaken a soda bottle and then unscrewed the cap? Increasing pressure pushes the bubbling liquid upward and soda explodes out the top! Well, deep in the earth, hot, liquid rock can build up pressure—like a shaken soda bottle. When it explodes onto Earth's surface, we're seeing a volcano in action!

Some volcanoes spew hot lava that can burn houses, trees, and cars. Some also release a lot of dust, gas, and debris into the air. Often this forms thick clouds. Eruptions can cause mudslides, avalanches (AV-uh-lanch-is), falling ash, and even floods. They can also trigger tsunamis (soo-NAH-meez), earthquakes, mudflows, and rockfalls.

Many volcanoes erupted during the global Flood. Today, Earth still has more than 1,500 active volcanoes, and at least 80 of them are under the oceans.

Did you know? The world's largest volcano, Mauna Loa in Hawaii, erupts about every five or six years.

Hawaiian Islands

- *The Hawaiian Islands are a row of islands in the Pacific Ocean.*
- *An underwater volcano formed them near the end of Noah's Flood.*
- *The seafloor slid sideways over a volcanic hot spot, so one hot spot formed several islands.*

The Hawaiian Islands are a row of small, beautiful spots of land in the middle of the Pacific Ocean. Where did this cluster of islands come from?

During the year of Noah's Flood, magma from deep in the earth rose up through the ocean's rocky floor. As it cooled, it formed volcanic cones on the seafloor north and west of Hawaii. Late in the Flood, the movement of tectonic plates beneath the seafloor slowed. This left more time for lava to build up. The volcanoes rose higher. They rose above the water to make the first Hawaiian Islands. Since the ocean plate beneath the islands was moving west, more islands formed over the same volcanic hot spot.

At the end of the Flood year, the plates slowed even more and the hot spot remained underneath the big island of Hawaii. Sometimes Hawaii's volcanoes erupt and spew out red-hot lava today!

Mount St. Helens

- *Mount St. Helens is an active volcano in Washington State.*
- *It erupted in 1980 and again in 1982.*
- *Big events can change Earth in a short amount of time.*

When Mount St. Helens shot steam into the air in March 1980, people flocked to Washington State. They wanted to watch the volcano erupt. But Mount St. Helens didn't blow up—it blew out the side!

Lava, ash, and rock traveled at over 300 miles per hour. Some ash shot about 15 miles into the air—higher than airplanes fly! Ash covered an area the size of West Virginia, and the event destroyed 230 square miles of land.

The second eruption in 1982 melted a lot of the ice at Mount St. Helens' crater top. Millions of gallons of water flowed down the mountain. This fast-moving water cut a canyon system that looked like a little Grand Canyon.

Mount St. Helens showed that big, messy events can cut canyons, move sand and soil, and change the look of the land in a short period of time. Studying what happened at Mount St. Helens helps us understand what happened to our planet during the global Flood described in the Bible.

Shaky Ground: Earthquakes

- *Earthquakes happen when Earth's tectonic plates rub together.*
- *The plates are like puzzle pieces floating on Earth's mantle.*
- *Big earthquakes can destroy cities.*

Rumble, rumble....Did you feel that?

Earthquakes happen all the time. Some are too small to notice. Others shake the ground and destroy cities! Around 500,000 earthquakes happen every year, but people only feel about 100,000 of them.

The ground we stand on is made of big slabs of rock that fit together like puzzle pieces. Sometimes those puzzle pieces, known as tectonic plates, move around and rub together. This creates earthquakes!

Earthquakes tend to happen in groups. When one earthquake rumbles, this can move other plates around, causing other earthquakes. Scientists can't predict these events, but they tell us where to watch for earthquakes.

Did you know? Earthquakes and underwater volcanoes can cause enormous, destructive waves called tsunamis (soo-NAH-meez).

Earth Rocks!

- *Earth has three types of rock: igneous, sedimentary, and metamorphic.*
- *Volcanoes and moving water can help form rocks.*
- *High heat and pressure can turn one kind of rock into another kind.*

We live on one rockin' planet! It's mostly made of rock and covered with rock layers. All of Earth's rocks can be put into one of three categories: igneous (IG-nee-us), sedimentary (SEH-duh-MEN-tuh-ree), and metamorphic (MEH-tuh-MOR-fik).

When magma from deep in the earth erupts as lava from a volcano, it cools and hardens into **igneous rock**. This material makes up most of Earth's crust.

Sedimentary rocks form when water washes together a lot of sand-like material that later hardens. Thanks to the global Flood, most geologic layers that cover Earth's crust are made up of sedimentary rocks.

High heat and pressure can turn sedimentary or igneous rock into **metamorphic rock**. For example, limestone is a sedimentary rock made of tiny grains cemented together. With high heat and pressure, limestone changes into marble. Marble is one kind of metamorphic rock.

Igneous

Metamorphic

Sedimentary

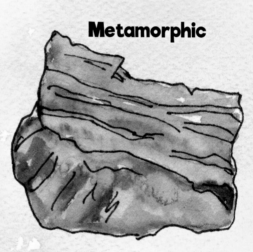

Explaining Earth's Rock Layers

- *Huge, multicolor rock layers cover Earth's continents.*
- *The Flood formed most of these rock layers in a short time.*
- *Scientists only observe small rock layers forming today.*

Today, we see some creeks deposit grains of sand and clay into thin layers. But did you know that huge rock layers cover whole continents? How did that happen? Some scientists say rock layers formed from everyday processes like river floods and slow changes in sea level over billions of years. But river floods never wash over whole continents!

If lots of time passed between the formation of each rock layer, we should see river cuts and animal burrows carved into each one. But we don't because there was no time for this to happen before the next layer was put down. So, Earth's rocky surface looks like a stack of flat, colorful pancakes that formed fast.

Noah's Flood best explains this. In only one year, the Flood's waters laid down huge amounts of sediment that turned to the gray, yellow, red, and brown rock layers that cover entire continents today.

Carving Caves

- *Caves are open spaces underground.*
- *They formed when water carried away limestone.*
- *Cave features can form much quicker than you might think.*

Caves run dark and deep beneath the ground. Most caves form when water carves openings in limestone rocks. Water carries a weak acid that breaks down the limestone. Around the end of Noah's Flood, plenty of fast-flowing water would have hollowed out huge caves in a hurry.

After a cave forms, dripping mineral water inside the cave can make weird and beautiful rock shapes called speleothems (SPEE-lee-uh-thems). Icicle-shaped speleothems form from cave roofs. They're called stalactites (stuh-LAK-tites). Cone-shaped speleothems form on cave floors. They're called stalagmites (stuh-LAG-mites).

Some stalactites several feet long formed in fewer than 50 years. Stalagmites have covered man-made cans and even bats. Caves and their speleothems can form fast with the right chemistry and water flow.

Did you know? Ice covered northern France after the global Flood. Ancient people started living in nearby caves and painted pictures of animals, people, and symbols on the walls.

Majestic Mountains

- *During and after the Flood, Earth's tectonic plates slammed into each other.*
- *This movement formed mountain ranges.*
- *Mountain ranges show evidence of the Flood.*

Have you ever driven through mountains? As your car rounds steep cliffs, you can look out over trees and rivers. As you go higher, the air turns chilly and snow blankets the rocky peaks. Don't you wonder how these mountains got here?

Scientists who don't believe the Bible think mountains formed gradually over millions of years. But creation scientists think mountains were formed by the global Flood. During the Flood, Earth's moving slabs of rock—known as tectonic plates—slammed into each other. Where they collided, the land pushed up. Floodwater runoff quickly carved the valleys and left the mountains behind.

Most of the world's mountains are made up of sediment normally found on ocean floors. And these sediments are full of sea creature fossils. How did the ocean floor and its creatures end up on mountaintops? A worldwide flood is the best answer.

Slippery Slopes

- *Harsh conditions wear away mountains and cause rockslides.*
- *The rocks pile up at the bottom of the mountains and create talus slopes.*
- *Earth has only thousands of years' worth of talus slopes.*

Have you ever seen rocks tumble down a mountain or canyon? Heat, cold, water, ice, and wind wear away mountains and cause these rockslides. Earthquakes and lightning can also cause them.

These rocks pile up at the bottom of the mountain or canyon over time. Scientists call these angled rock piles talus slopes. If talus slopes were forming over millions of years, there should be more of them all over the world. And most of the steep mountains we see shouldn't be steep any longer!

But scientists only find the number of talus slopes we would expect after thousands of years of rockslides, not millions. Talus slopes fit well with the Bible's 6,000-year timeline for the age of the earth.

Did you know? The breaking down of Earth's physical features—like wind wearing away mountains—can be slowed down, but it can't be stopped!

Digging for Diamonds

- *Natural diamonds form deep in the earth.*
- *They shoot up to Earth's surface through volcano-like pipes.*
- *Since diamonds still contain radiocarbon, they must have formed recently.*

You've probably seen diamonds—sparkling stones cut into perfect shapes. But these beautiful stones don't start out that way. They look dirty and lumpy when miners find them. It takes an expert to cut them and make them look nice.

How do diamonds form? Good question! Scientists can make diamonds in just a few weeks with the right mix of chemicals and some pressure. Natural diamonds form over 100 miles below the ground. They shoot through Earth's crust up long, volcano-like pipes.

Radiocarbon is often found in diamonds, and it can't last more than thousands of years. This means diamonds formed recently. Why does this matter? Evolutionists say diamonds formed billions of years ago. So, these dazzling stones must be younger than many scientists think. Scientists who believe the Bible know that young diamonds fit the Bible's 6,000-year timeline for Earth history very well.

Awesome Oceans

- *God gathered the waters into one big ocean on Day 3.*
- *The Flood formed newer and deeper oceans.*
- *Some ocean creatures are huge!*

If you've ever been to the ocean, you know it's big. Waves of water stretch farther than you can see. Like outer space, the ocean can make us feel very small.

God gathered the waters into one big ocean on Day 3 of the creation week. But as the waters went down after the global Flood, the ocean floor sank. This formed new and much deeper ocean basins. They still hold water from the Flood today.

Next time you're in an airplane, look out the window. See how far down the ground is? That's how deep some parts of the ocean are.

Scientists think that at least half of all life on Earth lives in the ocean. Some sea creatures can grow really big. Colossal squids are longer than school buses! The Japanese spider crab is bigger than your kitchen table. And some fish are longer than your living room.

Did you know? Scientists still have a lot to learn about Earth's oceans. They have better maps of Mars than of our ocean floors.

Turning the Tides

- *The moon pulls on Earth and its oceans.*
- *This causes the sea level to go up and down.*
- *These tides keep our oceans fresh and bring food to many creatures.*

If you ever go to the beach, you may notice the water slowly creeping farther onto the sand. Later on, it will slip back where it was. That's because the moon's gravity makes Earth's oceans go up and down each day. We call these changes tides.

As the moon orbits around Earth, it pulls the ocean toward it. The waters bunch up beneath the moon. As the water bulge slides around Earth, tides rise and fall. How much the sea level changes during high and low tides depends on the positions of the sun and moon, and Earth's rotation.

The ever-changing tides bring food and nutrients to sea creatures and plants. These plants release oxygen into the air for all of us to breathe. Praise the Lord, who daily turns the tides!

Did you know? Surfers study tides to figure out when and where they can ride the best waves.

Life-Giving Water

- *Water is necessary for life.*
- *Water covers 70% of Earth.*
- *Water shows God's care for Earth's creatures.*

Every living thing needs water—including people. We drink water, wash with water—even our bodies are made of over 60% water. Thankfully, God created Earth with lots of it!

Water covers about 70% of Earth's surface, and most of it is in the oceans. It comes in three forms: solid, liquid, and gas. Snow and ice are solid water. We see liquid water in lakes, oceans, and rivers…and it comes out of our faucets at home. Water in gas form is called water vapor. It floats in the air and forms clouds high in the sky.

Water stays in liquid form at Earth's typical surface temperatures. When water gets cold enough to freeze, the ice floats to the surface and protects the water beneath it from freezing. If this didn't happen, all of the water on our planet would turn into ice, making life impossible. Earth's huge supply of liquid water shows God's special care for people and animals.

Did you know? Scientists found evidence that another ocean's worth of water rests deep inside the earth.

The Water Cycle

- *Earth reuses the same water over and over again.*
- *Water moves in a cycle over land, sky, and sea.*
- *God uses the water cycle to provide for Earth's creatures.*

Did you know that Earth has recycled the same water over thousands of years? Water moves all around our world, passing over land, through the seas, and even into the sky. We call this the water cycle.

Rain and melting snow or ice run into rivers and streams and flow into lakes and oceans. Then, sunlight and wind cause the water to evaporate. The invisible vapors form clumps of tiny droplets in the sky called clouds. When the clouds get heavy, the water returns to Earth's surface as rain or snow. And the cycle starts all over again.

God created the water cycle to provide for all living things. This process turns salt water into fresh water that we can drink. It also helps plants to grow food we can eat. Our Creator is like a gardener who sprinkles water all around— giving animals and people just what we need to live.

Did you know? Since Earth recycles water, you could be drinking the same drops that a dinosaur sipped from a pond thousands of years ago.

Water Works

- *Earth's water supply isn't always balanced.*
- *Some lands have a lack of water—a drought.*
- *Other lands have too much water—a flood.*

We don't live in the perfect world God first created. So, unfortunately, Earth's water supply isn't always balanced. No rain for months or years is called a drought. Droughts are hard to predict because weather always changes. Long droughts can kill plants, trees, and animals.

Too much water can also be a problem. Floods are usually caused by many days of rain. Rivers fill up too fast and overflow their banks. Floods destroy houses, towns, and bridges. They can also bring disease.

Water changes the land it flows over. Weathering happens when water wears away Earth's surface, breaking down rock and dirt. Erosion happens when water or wind carries rock particles to another location. Grand Canyon is a great example of both weathering and erosion. Lots of water from the global Flood poured through the area, carving out the huge canyon and sweeping the rock and dirt toward the Pacific Ocean.

Raging Rivers

- *Rivers are long bodies of flowing water.*
- *Rivers support human and animal life.*
- *Most major cities are built along rivers.*

"Row, row, row your boat, gently down the stream…." Whatever you call them—streams, rivers, or creeks— long bodies of flowing water have been really important throughout human history.

Rivers flow all over the earth. Each one has three parts: a source, a course, and a mouth. The source is usually in the mountains, or some higher area, flowing down. The mouth is where the river ends and feeds into an ocean, sea, lake, or another river. And the course of the river is everything else in between. Water in rivers comes from rain or melting snow and ice.

Most major cities are built along rivers. This is because rivers are very useful for humans. They provide water for drinking and for growing crops like corn and wheat. We also use rivers for transportation so we can travel and trade with each other. Many animals live in and around rivers, including fish, frogs, beavers, and turtles.

Did you know? A river flowed through the Garden of Eden (Genesis 2:10).

Paths of the Sea

- *Biblical thinking led to a discovery about Earth's oceans.*
- *Ocean water flows in paths throughout the world.*
- *Water temperature and wind affect these paths.*

About 200 years ago, a man named Matthew Fontaine Maury was reading the Bible. Psalm 8 talked about "the fish of the sea that pass through the paths of the seas." Matthew wanted to know what it meant.

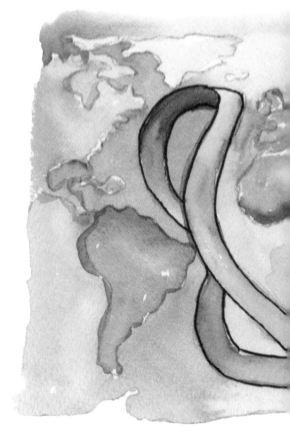

He and his team mapped the movements of thousands of floating objects called buoys. He realized that the water flows like paths through the seas. Whales swimming in the Pacific Ocean were later found in the Atlantic Ocean. So Matthew knew they must be traveling a path from one sea to the other. Even today, ships take advantage of these paths to save costs.

We now know that ocean paths tend to flow in circles caused by differences in water temperature, water saltiness, and wind. You'll notice these circular patterns in the map of ocean currents on these pages. Matthew's story shows how biblical thinking can lead to a better understanding of our world.

Climate and Weather

- *Climate is what the weather is like in one place over a long time.*
- *Earth has three climate zones: tropical, temperate, and polar.*
- *Earth's tilt causes different climates in different parts of the world.*

Weather is the temperature, sunshine, wind, and rain at one place at one time. But climate is the weather pattern over a year—or even longer.

Does your town get several feet of snow every winter, or do you live where you can wear shorts year-round? Thinking about these questions can help you figure out what climate you live in. Or, you can just check out the map on the opposite page!

Earth is tilted as it orbits the sun, so different parts of the planet get different amounts of sunshine throughout the year. This is why we have three climate zones: tropical, temperate, and polar. The tropical zone gets the most sunshine, so it's the warmest. The polar zones at the North and South Poles get much less sunshine, so they are the coldest. And the two temperate zones get changing amounts of sunshine, so they experience four distinct seasons. Which zone do you live in?

Did you know? The climate where you live is also affected by nearby oceans or lakes and whether or not you live in the mountains.

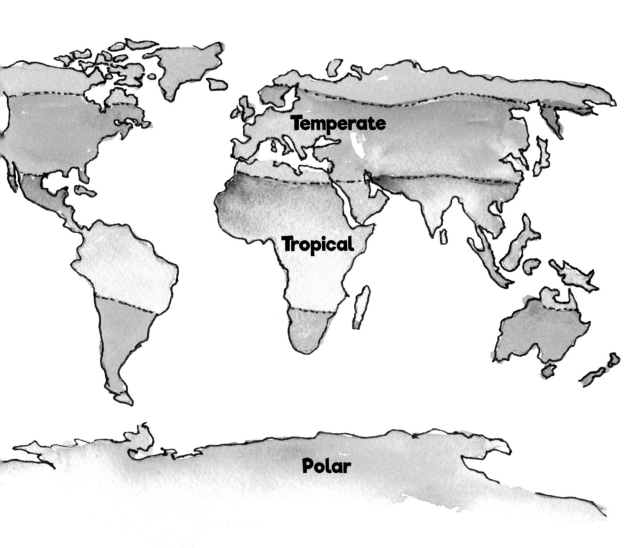

Temperate

Tropical

Polar

A Blanket of Air

- *The atmosphere is a thin blanket of gases that wraps around Earth.*
- *It has five main layers.*
- *It's necessary for life to exist.*

God wrapped Earth in a thin, cozy blanket of gases called the atmosphere. It's mostly made up of nitrogen and oxygen, though tiny amounts of several other gases are in it as well. Without our atmosphere's five layers, life could not exist.

Sunlight heats Earth's surface and warms the atmosphere's lowest layer—the **troposphere**. The troposphere keeps Earth from getting too hot or cold. All the weather happens here, too. Flying on a jet? Try the **stratosphere**—it allows a smoother plane ride. This layer also protects us from harmful sunrays.

Most meteoroids burn up in the **mesosphere** before they can hit Earth. Satellites orbit in the **thermosphere**, and that's also where natural light shows called auroras appear. In the highest layer, the **exosphere**, the air is so thin that it's almost like being in outer space.

Did you know? The sky looks blue because of the way water in Earth's atmosphere scatters sunlight.

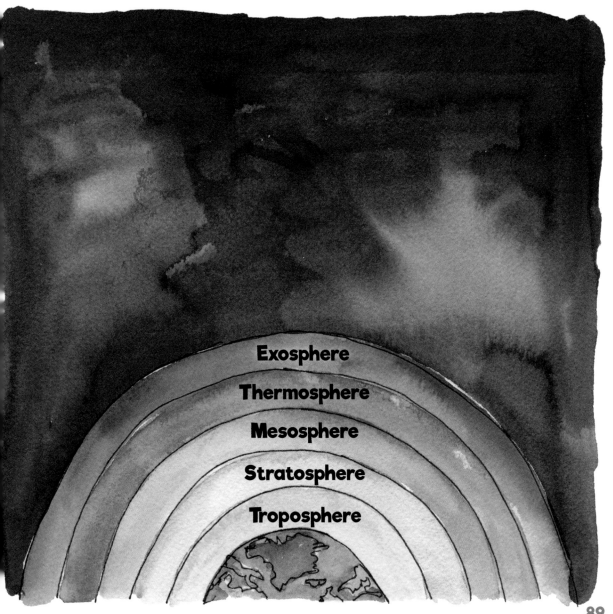

Spotting Clouds

- *Clouds are made of water droplets in the sky.*
- *There are five cloud kinds.*
- *Clouds impact Earth's weather.*

Do you ever search the sky for pictures in the clouds? If you do, you're what's known as a cloudspotter. But how do these interesting clouds get there in the first place?

As the sun's heat warms our oceans, lakes, and streams, the water rises in a gas form known as water vapor. When the vapor cools high in the sky, it clumps into tiny water droplets and forms clouds.

God created five different kinds of clouds that float in the sky. Some of them give clues for today's weather. Which kinds do you see?

Cirrus clouds are thin and wispy. They float so high that they are made of ice crystals rather than water droplets. **Stratus** clouds are low rainclouds and often form in flat layers. Thick ones bring rain or snow. **Stratocumulus** clouds look like ripples. **Cumulus** clouds are puffy and small. They tell us we'll have good weather! **Cumulonimbus** clouds are thunderclouds. Heavy and dense, these clouds can bring us hail and even tornadoes!

Blowing in the Wind

- *Wind is moving air.*
- *Wind can be a gentle breeze or a powerful hurricane.*
- *Wind is mostly caused by the sun warming Earth.*

We can't see the wind, but we see its effects when leaves rustle in the trees. What causes the wind to blow? A combination of hot and cold air.

The sun gives Earth different amounts of heat in different places. Warm air weighs less than cold air, so the warm air rises. Then, cold air fills in where the warm air used to be. This movement makes the wind blow. Cold air and warm air push against each other. The bigger the temperature difference is between them, the faster the wind.

As Earth rotates, it spins underneath the atmosphere. This causes the air to spin in complex spiral patterns and helps give us our weather patterns.

Powerful wind can cause damage, but moderate wind is useful. For many years, people harnessed wind to sail their ships, and they used windmills to grind grain and pump water from underground wells. Wind carries seeds and pollen to help plants grow. It also dries wet ground and can be used to make electricity.

Changing Seasons

- *Seasons are a cycle of weather changes throughout the year.*
- *Earth's changing position from the sun causes seasons.*
- *Seasons are opposite in the northern and southern parts of the earth.*

Each season comes with its own beauty, wonder, and purpose. Seasons happen because our planet's position around the sun changes throughout the year. This impacts the temperature and amount of daylight different areas receive.

Seasons are opposite in the northern and southern parts of the world. When it's winter in the United States and Europe, it's summer in South America and Australia. Places near Earth's equator have warm weather all year since the amount of sunlight they receive doesn't change much.

What's your favorite season? In fall, leaves transform into dazzling shades of orange, brown, and red. In winter, snow often blankets the earth. In spring, trees grow new leaves and flowers bloom. And long, hot summer days are great for cooling off in the pool.

If God had given Earth only one season, what would you miss?

To everything there is a season, a time for every purpose under heaven. (Ecclesiastes 3:1)

Icy Artwork

- *Snowflakes form when drops of water freeze and stick together.*
- *Air temperature and humidity determine a snowflake's shape.*
- *Most likely, no two snowflakes are exactly the same.*

Snow falls to Earth in a blur of winter white. But if you look closely, you'll find that each flake is a work of art. How do snowflakes form? In cold weather, drops of water in the atmosphere freeze into ice crystals. The crystals stick together to form a snowflake. On average, each snowflake is made up of about 200 ice crystals.

At the time a snowflake forms, the air temperature and humidity determine its shape. Colder air makes a snowflake with sharper tips. When the air is warmer, the ice crystals form more slowly and develop rounder shapes. Less moisture in the air makes simple shapes, and fancier snowflakes form when humidity is high.

Snowflakes grow and change even more as they fall from the sky. Ice crystals form, clump together, and fall under different conditions. So, each snowflake is unique, even though billions of them can fall in one snowstorm.

In the Eye of the Storm

- *Hurricanes are large, swirling storms with powerful winds.*
- *A hurricane's calmest part is its "eye" near the center.*
- *When hurricanes hit land, they can flood and damage cities.*

What's the recipe for a hurricane? Well, you need three ingredients: warm water, a giant ocean, and high air pressure above low-pressure thunderstorms.

In the summer and fall, thunderstorms travel west across Africa and form into large storms over the Atlantic Ocean. These low-pressure storms grow over the warm ocean because they pick up more and more water droplets as they swirl around.

When the winds get up to 74 miles per hour, the storm becomes a hurricane. Hurricanes have a calmer area near their center called the "eye." Can you find the hurricane's eye on the next page? The most powerful part of a hurricane is the border around the eye called the eye wall. There, the wind can blow over 200 miles per hour. When hurricanes hit land, they can flood cities and neighborhoods. Their powerful winds can knock down trees and houses.

Wild Weather

- *Sometimes Earth has extreme weather.*
- *Lightning, hail, and tornadoes can be dangerous.*
- *Stay calm and remember these safety tips.*

What causes wild weather, and what can we do about it?

Colliding particles inside a thundercloud create an electric charge. When it meets other charges on the ground, **lightning** strikes. Thunder is the sound of air collapsing afterward. Though it's rare, lightning can strike people! Be sure to stay away from open fields, metal objects, trees, and swimming pools during a thunderstorm.

When powerful, upward-moving winds freeze water in mid-air, it falls to the ground as icy chunks called **hail**. Most hailstones are tiny, but some grow to the size of a softball or larger. Large hail can cause serious injury to people and animals. Enter a sturdy building to escape it.

Tornadoes are violently twisting funnels of air. When one stretches from a storm cloud to the ground, it destroys everything in its path. Scientists can identify the right weather conditions for tornadoes, but they don't fully understand why they form. If you know a tornado is nearby, find shelter immediately and stay away from windows.

When Lightning Strikes

- *Lightning is a bright flash of electricity during a thunderstorm.*
- *Lightning strikes can turn rock, sand, or clay into glass.*
- *Earth is young, and that may explain why these kinds of glass are so rare.*

You might have a fever when you're sick, but it usually won't get much hotter than 102°F. How hot is lightning? More than 50,000°F!

When a bolt of lightning hits the ground, the intense heat can melt rock, sand, or clay into a funny-looking clump. Geologists call these rock-like clumps fulgurites (FULL-ger-ites).

Fulgurites come in many sizes and shapes depending on the kind of ground the lightning strikes. Some look like balls of jagged glass, while others resemble twisted sections of pipe. Some could be mistaken for a beautiful piece of ocean coral.

Scientists estimate that fulgurites are formed somewhere on Earth every second of every day. Since Earth is only thousands of years old like the Bible's timeline indicates, that could explain why these fascinating formations are so rare.

Did you know? Some people use fulgurites to make beautiful jewelry.

Climate Change

- *Some people think the burning of fossil fuels is causing dangerous global warming.*
- *History shows natural warming and cooling cycles.*
- *Both science and Scripture reassure us that Earth's climate will remain stable.*

Cars, planes, and factories create a lot of smoke and release a gas called carbon dioxide. Some people think carbon dioxide floats into the sky and traps heat, warming the earth. They're afraid Earth is getting so hot that glaciers will melt, sea levels will rise, and life won't survive.

Before we all panic, let's look at the science. Humans started building factories about 150 years ago. If factories were to blame for producing carbon dioxide and heating Earth, then we should see temperatures much higher today than ever before. But in the Middle Ages, Earth was much warmer. Then it cooled. Now, it's warming up again.

These warming and cooling cycles seem to be natural changes in our climate. Some scientists think they're caused by activity on the sun. Whatever the reason, we don't need to fear that our planet's systems will get out of control. God promised that Earth would keep its normal cycles and seasons as long as it exists (Genesis 8:22).

Discovering Deserts

- *Deserts get very little rain.*
- *Many desert creatures sleep during the day and are awake at night.*
- *The Antarctic Desert is the largest on Earth.*

Think about the hottest summer days you've known. Times when the sun beat down on you, sweat dripped from your face, and you couldn't seem to guzzle enough water to satisfy your thirst. Life in most deserts is like that, but a lot hotter and drier, and every day holds more of the same.

God created animals to adjust to different environments. Some of them thrive in the desert. Most sleep during the day and come out at night. After sunset in the Sonoran Desert, kangaroo rats scurry, coyotes hunt, owls soar, and mule deer feed on shrubs and grasses.

Deserts have extremely low rainfall. The Antarctic Desert has snow and ice, but it gets very little rain. It can be just as difficult to live there as it is in deserts with sand.

Did you know? The desert is usually pretty dry, but when it rains, it pours! The giant saguaro cactus can hold eight tons of water after a good downpour.

Life in the Rainforest

- *Rainforests provide homes for many animals.*
- *Rainforest water impacts the world's weather.*
- *God designed helpful relationships between rainforest plants and animals.*

Many of Earth's creatures live in a rainforest—from black-tailed deer and tree frogs to three-toed sloths and colorful macaws. There are two types of rainforest: tropical and temperate. Tropical rainforests are lush, warm, and rainy all year long. Temperate rainforests have less rain but cooler weather than tropical rainforests. They get the rest of their water from coastal fog.

Did you know that rainforests affect the world's weather patterns? Water that evaporates from rainforest trees can fall in other areas as rain.

God designed special relationships between rainforest plants and the creatures that eat them. The agouti is the only animal with teeth tough enough to crack the rainforest's Brazil nuts. Each nutshell contains more seeds than the agouti can eat. When it buries the leftovers, it often forgets where it put them—oops! But this is great for the rainforest because more Brazil nut trees sprout up from the buried seeds.

Thriving Coral Reefs

- *Coral is made up of animals called polyps.*
- *Large colonies of coral are called reefs.*
- *Coral reefs fit well with the Bible's creation account.*

Coral is made up of tube-like animals called polyps. God created them to come in beautiful shades of red, pink, green, blue, orange, purple, and white. Colonies of coral—called reefs—only cover a small part of the ocean floor. But they're home to sea urchins, sponges, sea stars, worms, fish, sharks, rays, lobster, shrimp, octopus, snails, and more.

Coral reefs require special conditions

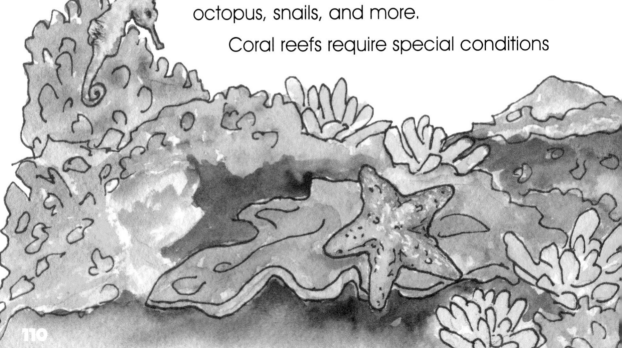

to form. They need clear and shallow water, sunny weather, and calcium. They grow in areas with lots of waves because the waves bring in food, nutrients, and oxygen.

Reefs soften the impact of strong waves and storms on land. And animals that live in reefs provide people with food.

Some scientists think huge coral reefs disprove the Bible's 6,000-year timeline. They say large reefs took more than 100,000 years to form. But studies show that coral can grow at very different rates depending on water temperature, sunlight, and other changing factors. Big and beautiful coral reefs fit just fine with the Bible's creation account in Genesis.

Ecosystems and Chores

- *A "household" of animals and plants is called an ecosystem.*
- *God gave each member of the ecosystem a special job.*
- *Ecosystem members work together to keep a healthy balance.*

Do you have chores at home? Maybe after a family dinner you pick up the dishes while someone else washes them. Everyone works together to help the household run. Well, the world has places like our households called ecosystems. Animals, plants, fungi, and bacteria in the ecosystem work together to live and thrive.

Each ecosystem has three main "chores": producing (making), consuming (eating or using), and decomposing (breaking down). Plants make food. Animals eat the plants—or each other. Fungi and bacteria break down dead plants and animals until they turn into dirt so the plants can make more food.

Balance keeps an ecosystem going. Plant-eaters keep plants from crowding their space and using up the earth's nutrients (food). Meat-eaters keep the plant-eaters from eating all the plants. And decomposers keep the area clean and return nutrients to the earth. Even in a fallen creation, we see God's wisdom.

Teeny-Tiny Life

- *Microorganisms are super-small living things.*
- *They live and thrive all over Earth.*
- *They keep creatures and their environments healthy.*

What's the smallest living thing you've seen? Maybe an ant or a doodlebug. Some living things are so small that you can only see them with a microscope. They're called microorganisms (MY-krow-OR-gan-izms). But let's call them microbes for short!

Cells are the building blocks of living things. Plants and most animals are made of many cells working together. But many microbes are only made up of a single cell. And they can be found all over Earth—deserts, mountains, forests, rivers, hot deep-sea vents, and even Antarctica! Some are so tough that they can survive outer space and harmful radiation.

Microbes like bacteria and fungi keep our world healthy. They clear away dead animals and plant remains. They also make many nutrients that help plants grow. Since our world is under the curse of sin, some kinds of bacteria can make you sick. But your gut uses good kinds of bacteria to help digest food. God created teeny-tiny life to do helpful work.

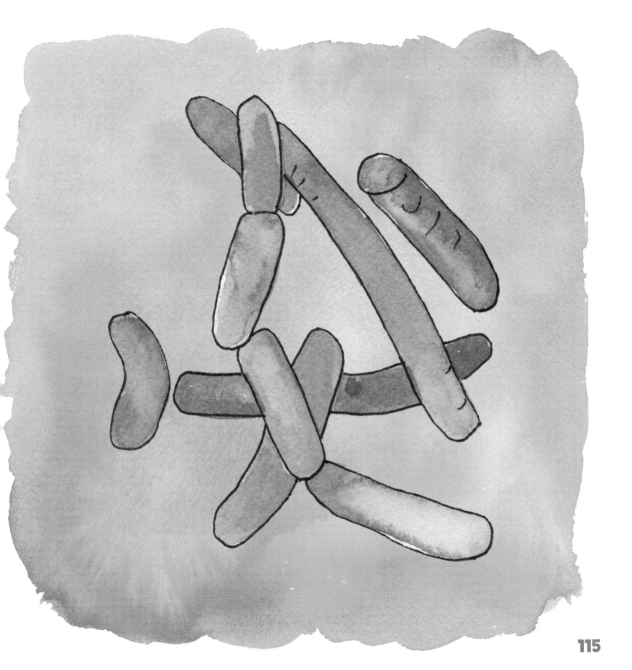

Goldilocks Planet

- *Earth is the only place where we know life exists.*
- *Earth is the only known "water planet." Seventy percent of its surface is covered by water.*
- *Earth is perfectly designed for life.*

God made Earth "just right" for life. That's why it's called the Goldilocks planet!

Earth orbits the sun at just the right distance so the planet isn't too hot or cold. And Earth's axis tilts just right to cause seasons that change throughout the year. People use seasons to determine planting and harvesting cycles for food. We also benefit from a rain and snow system that continually waters the earth.

The moon softly lights our nights, and its gravity causes gentle tides that refresh our oceans. Our atmosphere contains oxygen for creatures to breathe, and it also protects us from some of the sun's harmful rays.

God took great care in making Earth the perfect place for people, animals, and plants to thrive. Can you think of other "just right" features that show His care?

Did you know? Life is found all over Earth, from the coldest places to the hottest, and from high mountains to deep oceans.

Earth's Young Sun

- *The sun supports life on Earth.*
- *If the sun were billions of years old, it would have started out too dim for life to exist.*
- *God created the sun only thousands of years ago.*

What's the closest star to Earth? Look outside in the daytime and you'll find the answer glowing above you! God created our sun to be the star that rules the day (Genesis 1:16). The sun appears to move across the sky, but it doesn't really. The sun stays at the center of our solar system, and Earth rotates and orbits around it. It gives Earth just the right amount of heat and light for plants, animals, and people to live. And it releases more energy every second than a billion cities could use in a year!

Like all stars, the sun is slowly growing brighter and warmer. If it were really billions of years old as some say, it would've started out much dimmer. This would have made Earth too cold for life to exist.

The Bible's timeline indicates God created our sun only about 6,000 years ago. He made it at just the right time, distance, and brightness to support life on Earth.

Did you know? The sun is so big that over a million earths could fit inside it!

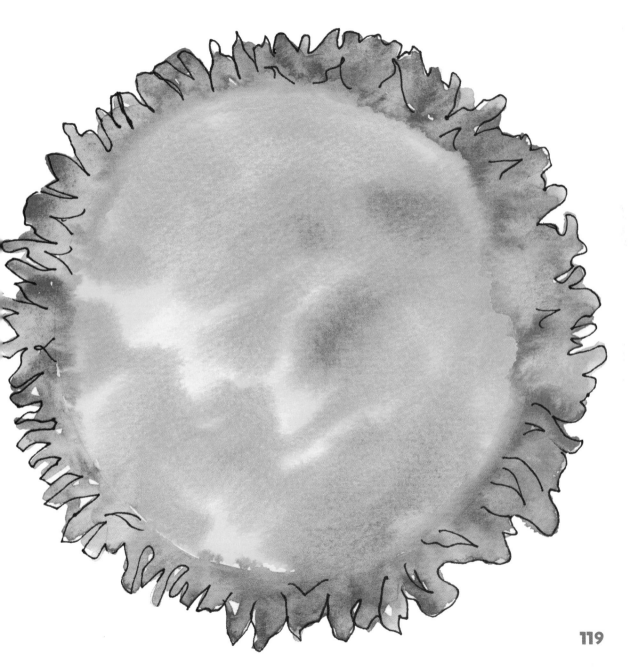

Amazing Auroras

- *Streaks of color light up the sky near Earth's poles.*
- *These auroras are caused by collisions in Earth's atmosphere.*
- *God created these beautiful light shows.*

If you like stunning views, you should see the night sky in Earth's Northern Hemisphere. Colorful streams of light called auroras shine in this part of the world. Sometimes we call them northern lights. Italian astronomer Galileo Galilei named the display aurora borealis (uh-ROAR-uh bor-ee-AL-is), which means "dawn of the north" in Latin.

Auroras happen when charged particles from the sun collide with gases in Earth's atmosphere near the North and South Poles. These collisions streak the sky in yellow, green, red, and blue!

God could have made them look ho-hum and boring. Instead, he made them colorful, dramatic, and totally awesome. Like so many things in our world, auroras remind us of our creative God.

Did you know? Auroras in the Southern Hemisphere are called aurora australis (uh-ROAR-uh aw-STRAY-lis), which means "dawn of the south."

View from Space

- *NASA astronaut Col. Jeffrey Williams viewed Earth from the International Space Station (ISS).*
- *He took many pictures of our planet from the ISS window.*
- *The ISS orbits Earth every 90 minutes.*

From the International Space Station (ISS), NASA astronaut Col. Jeffrey Williams got a big-picture view of Earth. He's rocketed into space four times and spent over 500 days orbiting high above our planet!

The ISS travels so fast that it circles Earth every 90 minutes. Col. Williams says he never gets tired of looking at our created world below. He's taken over 330,000 pictures from the ISS window!

What part of Earth does Col. Williams find most fascinating? He says,

> When you cross the Middle East, from orbit…you can see the entire biblical history…and the entire geography that Christ came and lived on as a man over 2,000 years ago…that gives deep and profound meaning to that view.

Did you know? The ISS has been orbiting Earth since 1998. If you know where to look, you can see it moving in the sky on a clear night.

Caring for Earth

- *God commanded humans to be His stewards on Earth.*
- *We should care for our planet and its creatures.*
- *The more we learn about Earth, the better caretakers we'll be.*

After He created them, God told Adam and Eve to "subdue" Earth and "have dominion" over its creatures (Genesis 1:28). This means to be a good steward, or manager, of our planet. He repeated and expanded this command when He spoke to Noah after the Flood (Genesis 9:1-7). Did you know it's still His plan for us today?

Being a steward of Earth doesn't mean we can treat our world however we want to. We are to care for creation in a responsible way that honors God. After all, Earth—and everything in it—is His.

What can you do to be a good steward today? You can keep Earth clean by picking up litter around your neighborhood. You can make the most of its resources by planting a garden or collecting items to recycle. And you can nurture Earth's animals by caring for your pet or putting a birdfeeder in your backyard.

What are some other ways you can show care for God's creation?

"Very Good" Again

- God created a perfect earth.
- God cursed Earth because of Adam's sin.
- Someday, God will make a new heavens and a new earth.

In this book, we've explored the many wonders of our home planet. But, in spite of its beauty, Earth still carries sin's curse. And its rocky layers full of fossils remind us of God's judgment upon sin during the global Flood. Will Earth ever go back to being "very good" again?

The Bible says that all of creation groans to be made new (Romans 8:21-22). At the right time, God will remake Earth, with all the scars of sin and death burned away (2 Peter 3:10).

People are living under sin's curse, too. We sin every time we disobey God's Word. Our hearts and bodies groan to be made new. The good news is that Jesus' death on the cross paid the price for our sin. If we put our trust in Him, we will live forever on the new earth with our Creator. Are you trusting in Him today?

> "For behold, I create new heavens and a new earth; and the former shall not be remembered or come to mind." (Isaiah 65:17)

Earth Resources for Deeper Discovery

For more detailed answers to your questions about Earth, visit ICR.org. Search for these articles, podcasts, and other news describing the latest scientific research. The books and DVDs are offered in our online store (ICR.org/store).

Articles

"What Was the Pre-Flood World Like?" Brian Thomas, Ph.D.

"NASA Earth Image Helps Answer Flood Question," Brian Thomas, Ph.D.

"The Genesis Flood and Evangelism," Jake Hebert, Ph.D.

"Was There an Ice Age?" Jake Hebert, Ph.D.

"Are Fossils Just Rocks Shaped Like Bones?" Brian Thomas, Ph.D.

"Is Young Earth Creation Crazy?" Brian Thomas, Ph.D.

"Don't Grand Canyon Rocks Showcase Deep Time?" Brian Thomas, Ph.D.

Podcasts

"Remembering Mount St. Helens," various scientists

"Why the Age of the Earth Matters," Jake Hebert, Ph.D.

"Day Three—Land and Seas," Henry M. Morris III, D. Min.

"The Flood of Fossils," Frank Sherwin, M.A.

Books

Big Plans for Henry

Dinosaurs: God's Mysterious Creatures

Animals by Design: Exploring Unique Creature Features

Space: God's Majestic Handiwork

Guide to Creation Basics

Guide to Dinosaurs

Guide to Animals

DVDs

That's a Fact 1 and *2*

Unlocking the Mysteries of Genesis

The Mighty, Wonderful Oceans

Uncovering the Truth about Dinosaurs

Glossary

Atmosphere Layer of gases surrounding a planet or other celestial body.

Aurora Streams of colored light that stretch across the sky at the North and South Poles. They're caused by interaction between solar particles, Earth's atmosphere, and Earth's magnetic field.

Axis The imaginary straight line that a rotating planet or moon spins around.

Climate The weather pattern in a specific region.

Climate change A change in climate patterns worldwide or in a specific region. Many scientists claim climate change is happening because of higher levels of carbon dioxide in Earth's atmosphere. But evidence shows that Earth's climate has cycled through warmer and cooler periods throughout its history.

Coral reef A rocky underwater ridge made up of animals called polyps and their hardened calcium carbonate shells.

Crust Earth's hard outer layer made of igneous, metamorphic, and sedimentary rock.

Current The movement of seawater in a specific direction because of factors like wind, gravity, Earth's rotation, and the temperature and saltiness of the water.

Desert A dry area of land known for low rainfall. A desert can be hot or cold.

Drought An extended time of unusually dry weather without rain or snow.

Earthquake A sudden shaking of the ground caused by movement in Earth's crust and/or volcanic activity.

Equator An imaginary line around Earth's middle that divides our planet into Northern and Southern Hemispheres.

Erosion The gradual wearing away of Earth's surface that moves bits of soil, rock, or other materials from one location to another. Erosion is caused by wind, water, or ice.

Evolution The theory that every creature formed from one original creature through chance and natural processes over millions of years.

Flood An event described in the Bible in which God covered the entire earth with water as punishment for the evil of its people.

Fossil The remains, imprint, or outline of an organism that lived in the distant past. Fossils can also include droppings, footprints, or preserved material. Fossilization of a plant or animal requires fast and deep burial beneath sediments like sand or clay.

Fossil fuel A natural energy source formed from the remains of buried organisms. Examples include oil, coal, or natural gas.

Fulgurite A crystal-like structure that forms when lightning strikes bits of sand, soil, or rock and fuses them together.

Glacier A large, heavy sheet of slow-moving ice.

Gravity An invisible force that pulls objects toward each other.

Hail Balls of frozen rain that grow and fall from clouds in small to large chunks.

Hemisphere One half of Earth. Our world can be divided into the Northern and Southern Hemispheres or Eastern and Western Hemispheres.

Hurricane A storm with very high winds that move violently in a circle. It forms over warm waters in tropical areas.

Ice Age A period after the Flood when some of Earth's water froze into broad sheets of ice on many land surfaces. The ice lasted for hundreds of years.

Igneous rock A rock formed from hardened magma or lava.

Inner core The solid center layer of Earth. Many scientists suspect it's made of nickel and iron.

Land bridges Strips of land that connected continents and allowed animals and people to migrate around the world during the Ice Age.

Lava Hot, liquid material that erupts from a volcano or opening in the earth. It becomes rock as it cools.

Magma Hot, liquid material beneath Earth's crust. Magma is called lava once it erupts above Earth's surface.

Magnetic field An invisible field produced by moving electrical charges that influence other moving charges.

Mantle The layer of Earth between the crust and core. It makes up over 80% of the Earth and is made of iron and magnesium-rich rocks.

Metamorphic rock A rock changed into another type of rock by very high heat and/or pressure. One example is marble, which forms when limestone is exposed to these extreme conditions.

Outer core Earth's liquid layer that surrounds the inner core. Scientists suspect it's made mostly of iron and nickel.

Radiocarbon A rare type of carbon atom that releases particles and radiation until it stabilizes into a nitrogen atom. It's used to estimate ages of recently deposited carbon-containing objects like wood, coal, or bone.

Rainforest A thick, lush forest known for high rainfall, tall trees, and a wide variety of plants and animals.

Season A time of year characterized by certain weather patterns and hours of daylight. Earth's four seasons—spring, summer, winter, and fall—are caused by Earth's changing position around the sun.

Sediment Small pieces of a solid material that settle at the bottom of a liquid. Sediments like sand and clay were put down in layers during the global Flood. Some sediment even forms from dissolved material like salt and limestone.

Sedimentary rock A rock formed when sediments like sand or clay are pressed and cemented together by strong forces and chemicals. Fast-moving water formed sedimentary layers during the global Flood that then hardened into rock.

Speleothem A natural structure that forms when water deposits minerals inside a cave.

Stalactite A mineral deposit that hangs from the roof of a cave.

Stalagmite A mineral deposit that extends from the floor of a cave.

Talus slope A sloped pile of broken rocks that collect at the bottom of a canyon or mountain.

Tectonic plates Massive slabs of rock that include Earth's crust and the very top layer of the mantle beneath it. The movement of these rocky slabs is called plate tectonics.

Tide The daily changes of sea level on the shore. They're caused by the gravity of the moon and sun.

Tornado A destructive, rotating funnel of wind that extends from a thunderstorm and touches the ground.

Tsunami A huge ocean wave that is typically caused by an under-water earthquake or volcanic eruption that moves the seafloor up or down.

Water cycle The continuous cycle of water in which water changes from water vapor in the atmosphere to liquid water on Earth and then back to water vapor again.

Index

Contributors

Susan Windsor is the graphic designer and illustrator for the Institute for Creation Research. Her favorite place on Earth is the beach, where she enjoys listening to the waves and seagulls.

Christy Hardy is a writer and editor for ICR. She loves camping by the lake and falling asleep to Earth's symphony of chirping crickets and rustling trees.

Truett Billups is a writer and editor for ICR. He enjoys snowboarding and camping. His favorite camping trip (so far) was to Palo Duro Canyon near Amarillo, Texas.

Michael Stamp is a writer and editor for ICR. He also writes Christian novels. His favorite place on Earth is Nova Scotia.

Brian Thomas is a dinosaur researcher, science writer, and speaker for ICR. He enjoys backpacking in the Rocky Mountains of Colorado.

Jayme Durant is the Director of Communications and Executive Editor for ICR. Cascading waterfalls, trees that reach to the blue skies, and one-of-a-kind snowflakes are some of her favorite things on Earth.

Thank You

A special thank you to Dr. Tim Clarey for his careful review of this book. His expertise in dinosaurs and geology proved invaluable as we sought to accurately represent ICR's commitment to solid science and biblical creation.

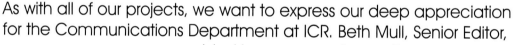

We'd also like to thank ICR's CEO Dr. Henry M. Morris III for his support throughout the development of this project, his biblical expertise, and his thorough review of this book.

As with all of our projects, we want to express our deep appreciation for the Communications Department at ICR. Beth Mull, Senior Editor, provided her seasoned expertise in reviewing and editing this book, and James Turner and Michael Hansen helped us develop ideas that would connect with kids for fun and learning.

About ICR

At the Institute for Creation Research, we want you to know God's Word can be trusted with everything it speaks about—from how and why we were made, to how the universe was formed, to how we can know God and receive all He has planned for us.

To build your faith, our scientists have spent decades researching how science supports what the Bible says. Our experts earned degrees in many different fields, including genetics, biotechnology, astronomy, astrophysics, physics, nuclear physics, zoology, geology, medicine, public health, theology, and engineering.

In addition to our books, we publish a monthly *Acts & Facts* magazine, a quarterly *Days of Praise* devotional, and loads of scientific articles online. We also produce DVD series on science and the Bible, and our scientists travel across the country to speak at events and share their findings. Our latest project is the ICR Discovery Center for Science & Earth History in Dallas, Texas—we'd love for you to visit! For more information on our ministry, go to our website, ICR.org.

Other Resources by ICR
SCIENCE FOR KIDS

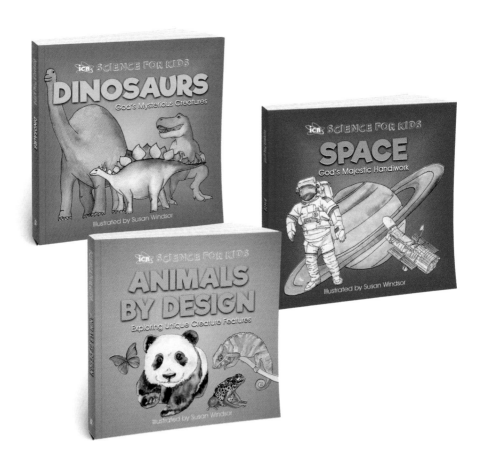

ICR.org/store

Family Resources from ICR

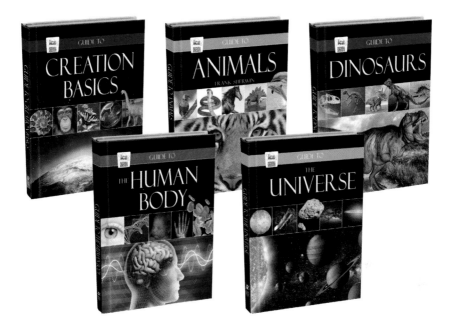

- Family-friendly, multi-age-level book series
- Packed with vivid illustrations
- Hardbound, full color
- Trusted biblical creation message